Gabriela Pasetto Falavigna
Sergio F. de Souza

Comparison between the TC2DFTPL and TCFOUR programmes

Gabriela Pasetto Falavigna
Sergio F. de Souza

Comparison between the TC2DFTPL and TCFOUR programmes

Terrain correction calculation

ScienciaScripts

Imprint
Any brand names and product names mentioned in this book are subject to trademark, brand or patent protection and are trademarks or registered trademarks of their respective holders. The use of brand names, product names, common names, trade names, product descriptions etc. even without a particular marking in this work is in no way to be construed to mean that such names may be regarded as unrestricted in respect of trademark and brand protection legislation and could thus be used by anyone.

Cover image: www.ingimage.com

This book is a translation from the original published under ISBN 978-613-9-71824-5.

Publisher:
Sciencia Scripts
is a trademark of
Dodo Books Indian Ocean Ltd. and OmniScriptum S.R.L publishing group

120 High Road, East Finchley, London, N2 9ED, United Kingdom
Str. Armeneasca 28/1, office 1, Chisinau MD-2012, Republic of Moldova, Europe
Printed at: see last page
ISBN: 978-620-7-95863-4

ACKNOWLEDGEMENT

This work is part of a wider project on the determination and evaluation of the geoid in the state of Rio Grande do Sul. This project has the support of CNPq in the form of a research project, through financial support received in the form of a PIBIC/CNPq-UFRGS Scientific Initiation Scholarship.

SUMMARY

CHAPTER 1

INTRODUCTION

Determining the geoid using gravimetric data involves solving the Geodesic Contour Value Problem (GCVP), which assumes two conditions (GEMAEL, 2002):

i.   gravimetric measurements must be made on the geoidal surface; and

ii.   there should be no masses outside the geoid.

The gravitational effect of topographic masses located above the geoid must be taken into account in various applications of Physical Geodesy, such as the calculation of gravity anomalies (free air anomaly, Bouguer anomaly, isostatic anomaly) and geoid undulations (MATOS, 2005).

The first PVCG condition is met through the free-air correction, which reduces the observed value of gravity to the geoid; and the second can be met through the Helmert condensation method (MATOS, 2005), whereby the topographic masses are removed and subsequently placed back inside the geoid, considering the external masses to have a specific density.

The free air anomaly is obtained by applying the free air correction, which takes into account the variation in gravity between the physical surface and the geoid surface, using the altitude of the point and disregarding the mass between these surfaces. The Bouguer anomaly is obtained from the Bouguer correction, which removes the effect of masses between the station and the geoid, assuming that the space between the physical surface and that of the geoid is filled uniformly with mass in the case of a station situated above the geoid and uniformly empty if the station is below the geoid. The isostatic anomaly takes into account the topo-isostatic correction, which also takes into account the heterogeneity of mass below the geoid, understood as compensation mass for the topography above the geoid (JAMUR et al., 2010).

Terrain correction is used in some of the reductions applied to the acceleration of gravity, such as the Bouguer anomaly and the Helmert reduction (GEMAEL, 2002). By applying terrain correction to gravimetric anomalies, the masses outside the geoid are vertically compressed beneath the geoid's surface, altering the Earth's gravitational potential. Currently, it has been possible to systematically calculate this correction using local or global Digital Terrain Models (DTMs) (JAMUR et al., 2010), as the altitudes are provided in grids of different sizes that allow the use of linear mass topographic models and rectangular prisms, through the use of the FFT (Fast Fourrier Transform) in the terrain correction formula.

In this work, the TC2DFTPL and TCFOUR programmes were used together with data from the Shuttle Radar Topographic Mission (SRTM) MDTs and the Advanced Spaceborne Thermal Emission and Reflection Radiometer (ASTER) Global Digital Elevation Model (GDEM) to calculate the terrain correction to be applied to the gravimetric observations of the state of Rio Grande do Sul.

CHAPTER 2

GROUND CORRECTION

Gravimetric observations made on the surface of the ground are strongly affected by topography, and their effect is greater the higher the elevation at which the measuring station is located.

Terrain correction (TC) aims to regularise the topographic surface, thus obtaining a topography defined by a plateau with the altitude of the measuring station and constant density, called the Bouguer Plateau. As the Bouguer Plateau has a constant thickness and is equivalent to the altitude of station P, masses can occur around this point which are not taken into account or which are removed without existing. As a result, the observed gravity value becomes smaller than the real one, meaning that corrections for the effect of terrain irregularities in relation to the station's altitude plane are always positive (LOBIANCO, 2005). In order to solve this problem and refine the Bouguer anomaly, it is necessary to add a component due to the topographic masses that were not taken into account above station P and a component to correct the mass incorrectly taken into account below this station (Figure 1) (MATOS, 2005). Therefore, the terrain correction, in flat approximations (FORSBERG, 1984), always assumes positive values, thus increasing the observed gravity value.

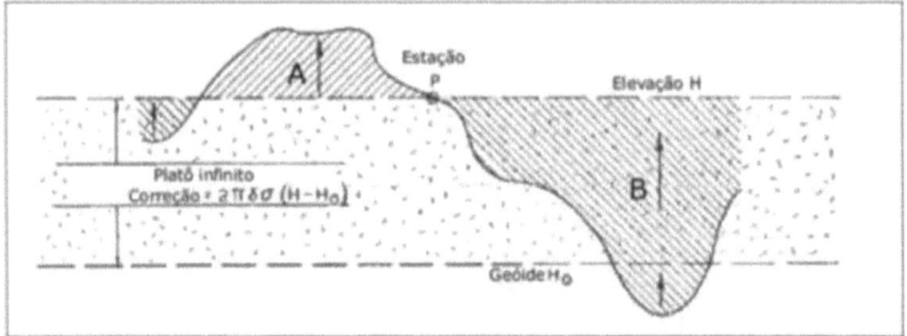

Figure 1 - Terrain correction. **Source:** Adapted from Matos (2005).

Hammer (1939) developed a system for assessing terrain correction (LOBIANCO, 2005). Hammer's method for terrain correction considers dividing the area around the station into zones and circular compartments (MATOS, 2005). The difference in altitude between the calculation point and each surrounding compartment is estimated, making it possible to calculate the effect of these masses on the station. The radii and number of compartments that each Hammer zone is divided into are tabulated. Thus, the terrain correction contribution for a point is given by the sum of the tabulated values for each zone (LOBIANCO, 2005). The curvature of the Earth is ignored in the Hammer Tables, as its effect is practically negligible within the area covered by the tables; only in the centre of zone M does the Earth's curvature begin to be considerable. The density of the surface material adopted in the Hammer Tables is 2.0 g/cm^3 ; for another density, the total terrain correction determined by the tables must be multiplied by the factor: density divided by two (LOBIANCO, 2005).The classic terrain correction formula is given by (MATOS, 2005):

$$c(x_p, y_p) = \frac{G\rho R^2}{2} \iint_E \frac{[h(x,y)-h(x_p,y_p)]^2}{l^3(x_p-x, y_p-y)} dx\, dy \qquad (1)$$

where G *is* Newton's gravitational constant;

ρ is the density of the topographic masses, assumed constant and equal to

2.67 g.cm^{-3} ;

R is the radius of a sphere approximated to the global geoid;

(x_p, y_p) is the coordinate of the calculation point;

(x, y) are the coordinates of the MDT points;

h is the altitude of the point above mean sea level;

E denotes the area of integration of the surface;

$l(x_p - x, y_p - y)$ is the *kernel* defined as the distance between the points (x_p, y_p) e (x, y):

$$l(x_p - x, y_p - y) = \left[(x_p - x)^2 + (y_p - y)^2\right]^{0,5} \qquad (2)$$

The Hammer method and the Equation generate identical results. The former, although efficient, is time-consuming and subject to human error, as well as being unfeasible for obtaining terrain correction for a large number of points, as is the case when calculating local regional models (LOBIANCO, 2005. A quick and automated alternative for obtaining terrain correction on small scales is to use programmes that evaluate the equation as a convolution integral on a regular grid using the fast Fourier transform (FFT-2D). This is made possible by the relationship:

$$F\{a * b\} = F\{a\}F\{b\} \qquad (3)$$

where $a * b$ *is* called the convolution of functions a *and* b, *and* F$\{a\}$ and $F\{b]$ are the Fourier transforms of a and b, respectively (MATOS, 2005). Equation 3 is applied to Equation 1, where the function a is the term $[h(x,y) - h(x_p, y_p)]^2$ and the function b corresponds to $l^{-3}(x_p - x, y_p - y).$.

The use of FFT-2D in the calculation of terrain correction considerably and effectively reduces the calculation time, the laborious process and the human errors involved in manipulating Hammer templates (LOBIANCO, 2005). Further details on the use of the FFT in the calculation of terrain effect can be found in Forsberg (1984).

Terrain correction takes on relatively small values. On a regional scale it can be neglected as long as the topography is flat or moderate. On a local scale, however, it needs to be considered, as the anomalies involved in this case are usually small. In mountainous areas, with altitudes of around 3,000 metres, terrain correction also needs to be considered, as in this situation it can take on values of around 50 mGal (LOBIANCO, 2005).

CHAPTER 3

METHODOLOGY

The study area (Figure 2) for this work is located in the southern part of Brazil and corresponds to the state of Rio Grande do Sul (RS). RS covers a total area of 281,730.223 km² (IBGE States, 2014) and lies between latitudes 33° 43'S and 27°05'S and longitudes 49°42'W and 57°40'W.

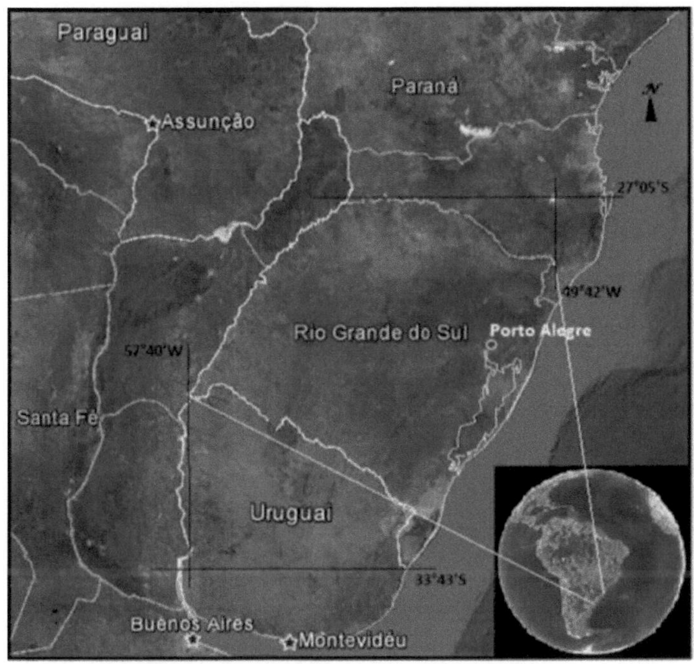

Figure 2 - Location of the study area.

3.1. DATA USED

The SRTM (Shuttle Radar Topographic Mission) is an international project led by the National Geospatial-Intelligence Agency (NGA) and the National Aeronautics and Space Administration (NASA) with the aim of producing a high-resolution digital topographic model of the Earth (NASA, 2013). This mission was carried out by the space shuttle Endeavour between

11 and 22 February 2000, covering around 80% of the Earth's surface (area between latitudes 60°N and 56°S), with a spatial resolution of one arc second (1"), which corresponds to approximately 30 metres. The altimetry data acquisition method used was radar interferometry, which consists of comparing two radar images taken from slightly different points to obtain the elevation. To carry out this procedure, the space shuttle Endeavour was equipped with a 60-metre mast and two receiving antennas (C-band and X-band), one of which was installed in the shuttle's cargo compartment and the other on the end of the mast that extended out of the spacecraft; this made it possible to acquire the data in the same orbit, thus guaranteeing the best quality (RODRIGUEZ et al., 2005). SRTM images have a resolution of 30 metres for the United States and 90 metres for other countries; its altimetric reference is the WGS84 ellipsoid and its terrestrial model is represented by the Earth Gravitational Model 1996 (EGM96). The SRTM data for Brazil can be downloaded free of charge from the UFRGS Ecology Centre website[1] .

An assessment of the quality of SRTM altimetry data for RS can be found in Lemos et al. (2004) and for Brazil and Argentina in Blitzkow et al. (2007). The preliminary results of the analysis of the quality of SRTM-derived altimetry data for an area within RS, obtained by Lemos et al. (2004), showed that for many applications SRTM can replace MDTs obtained from topographic maps at a scale of 1:250,000.

The Advanced Spaceborne Thermal Emission and Reflection Radiometer (ASTER) Global Digital Elevation Model (GDEM) was

---

developed jointly by Japan's Ministry of Economy, Trade and Industry (METI) and NASA. ASTER GDEM contributed to GEOSS (Global Observation System of Systems) and is available free of charge to users via download from the websites of Japan's ERSDAC (Earth Remote Sensing Data Analysis Centre) and LP DAAC *(NASA's Land Processes Distributed ActiNe). rcNe/e CeNter).* In this work, the ASTER data for the study area was acquired from the ERSDAC site[1] [2]

ASTER was built by METI and launched by NASA's Terra spacecraft in December 1999. Throughout its trajectory, it has stereoscopic capabilities using the near-infrared spectral band and a spatial resolution of 15 metres in the horizontal plane. It covered the area of the Earth's surface between latitudes 83°N and 83°S (22,600 1° x 1° areas - areas containing at least 0.01% of the Earth's surface were included); it is in GeoTIFF format, with lat/long geographic coordinates and a 1" arc grid (approximately 30 metres); its altimetric reference is WGS84 and its terrestrial model is represented by EGM96 (GISAT, 2014).

The methodology used for its production involved the automated processing of the entire ASTER scene archive (1.5 million scenes). Stereo-correlation was used to produce 1,264,118 individual ASTER DEM base scenes; cloud masking to remove bad pixels; merging of all selected DEM clouds, removing bad residual values and outliers; use of the average of the selected data to define the final pixel values and correction of residual anomalies before partitioning the data into 1° x 1° areas (GISAT, 2014).

Previous productions estimate (but do not guarantee) an accuracy for this global product of 20 metres with 95% confidence for vertical data and

30 metres with 95% confidence for horizontal data (GISAT, 2014).

Druzina (2007) compares the MDTs generated for the municipality of Porto Alegre/RS using different data (topographic maps at a scale of 1:50,000, ASTER and SRTM) with ground-truth MDTs. The statistical results of these evaluations showed that the ASTER MDT was the model with the lowest Mean Squared Error in a hilly area (6.522 m) and the second lowest Mean Squared Error in a flat area (4.626 m) and in a mixed area (6.698 m).

3.2. PROGRAMMES USED

The FORTRAN77 programme called TC2DFTPL, developed by Sideris (1985) and Li and Sideris (1994), uses Equation 1 as a convolution integral over a regular MDT grid and calculates the terrain correction via FFT-2D (MATOS, 2005). For the program to run, you need to enter: the name of the file containing the altitudes, the number of rows and columns and the resolutions, in kilometres, in x (dx_km) and y (dy_km) of the file. The information on the number of rows and columns and the resolutions must be placed on the first line of the altitudes file, separated by a space; then the programme must be told the order in which this information is arranged in the altitudes file. After the programme has processed the data, three output files are produced: the first is the terrain correction to be applied to the gravimetric data considering the topographic model as having the mass concentrated in a prism; the second is the terrain correction to be applied to the gravimetric data considering the linear mass topographic model; and the third file averages the two previous output files.

In this work, in order to calculate the terrain correction using the TC2DFTPL programme, the topographic model was considered to have its mass concentrated in a prism.

TCFOUR is an IAG (International Geoid School) programme and is part of the GRAVSOFT suite (JAMUR; de FREITAS, 2012). This programme uses two MDT grids with different resolutions to calculate the terrain correction. The first grid, called the

The detailed or high-resolution grid is the one that recovers most of the effect of the terrain on the magnitude of the gravity field due to its proximity to the calculation point, and is considered up to a radius of R0. The second grid is the so-called reference grid, which acts as a high-pass filter (JAMUR; de FREITAS, 2012). For the programme to run, you need to enter the names of the files containing the input data, in this order: detailed grid; reference grid; name of the output file; option of what you want to calculate (option 3 for terrain correction); radius (in km) to be taken into account in the calculation; number of rows and columns in the file. The TCFOUR programme calculates the terrain correction via FFT for a flat approximation (FORSBERG; TSCHERNING, 2008) and generates an output file.

In this work, the highest resolution grids used were 10 km, 5 km, 2 km and 1 km, extracted from the SRTM MDTs (with a spatial resolution of 90 m) and ASTER (with a spatial resolution of 30 m), using an R0 = 999 km. The resolution of the reference grids used was 20 km, also extracted from the SRTM and ASTER MDTs.

CHAPTER 4

CT CALCULATION USING THE TC2DFTPL AND TCFOUR PROGRAMMES

The terrain correction was calculated using the TC2DFTPL and TCFOUR programmes using data from the SRTM and ASTER MDTs at resolutions of 20 km, 10 km, 5 km, 2 km and 1 km. In these programmes, the TC was calculated for the SRTM and ASTER MDTs at resolutions of 10 km, 5 km, 2 km and 1 km. The 20 km resolution MDT was used as the reference grid in the TCFOUR programme.

In the TC2DFTPL programme, for the SRTM MDTs, it was necessary to divide the RS area only for the 1 km model. Thus, the RS area was divided into four parts. For the ASTER MDTs, it was necessary to divide the study area for the 2 km and 1 km models. Thus, the study area was divided into two and nine parts respectively.

In the TCFOUR programme, for the SRTM MDTs, it was not necessary to divide the RS area for any of the models. For the ASTER MDTs, however, it was necessary to divide the study area for the 1 km model. Thus, the RS area was divided into four parts.

After calculating the TC using the programmes, it can be seen that the TCFOUR programme showed an edge effect for the study area and for the necessary divisions of the study area. Unlike the TC2DFTPL programme, which showed no edge effect in any situation.

The division of the study area was necessary due to the limited data processing capacity of the programmes for calculating the Fourier transform in a single convolution. Table 1 shows for which MDT resolutions it was necessary to compartmentalise the study area.

Table 1 - Compartment requirements of the study area for calculating the TC.

Programme	MDT resolution	SRTM	ASTER
	10 kilometres	No	No
	5 kilometres	No	No
TC2DFTPL	2 kilometres	No	Yes
	1 kilometre	Yes	Yes
	10 kilometres	No	No
	5 kilometres	No	No
TCFOUR	2 kilometres	No	No
	1 kilometre	No	Yes

The spatial layout of the terrain correction calculated by the TC2DFTPL programme from the SRTM and ASTER MDTs at different resolutions can be seen in Figures 3 to 6 and Figures 7 to 10, respectively.

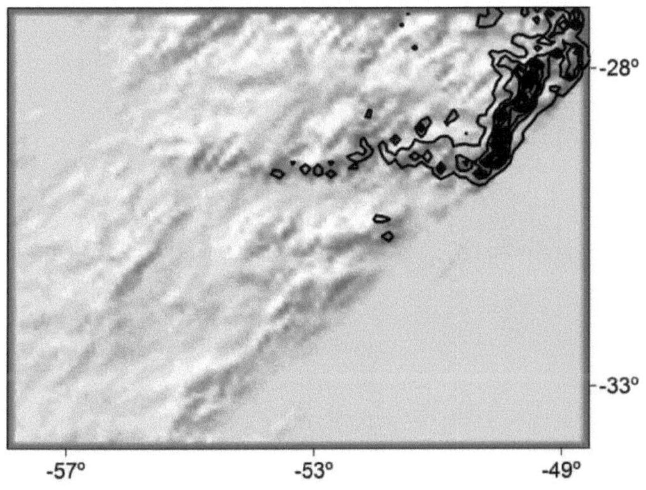

Figure 3 - Spatial layout of the TC calculated by TC2DFTPL for the SRTM MDT with 10 km resolution

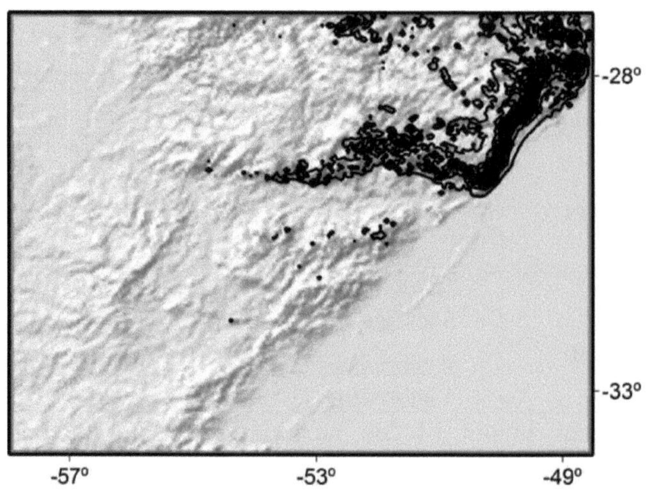

Figura 4   - Spatial layout of the TC calculated by TC2DFTPL for the SRTM MDT with 5 km resolution

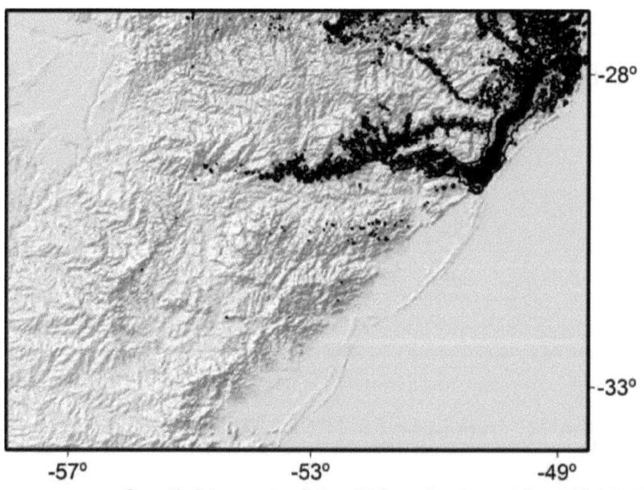

Figura 5   - Spatial layout of the TC calculated by TC2DFTPL for the SRTM MDT with 2 km resolution

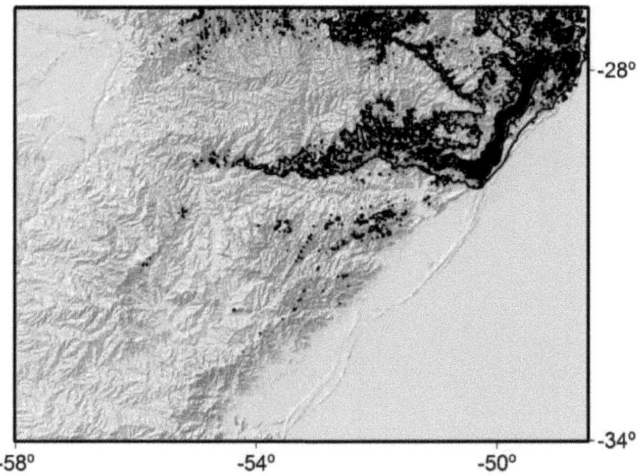

Figura 6 - Spatial layout of the TC calculated by TC2DFTPL for the SRTM MDT with 1 km resolution

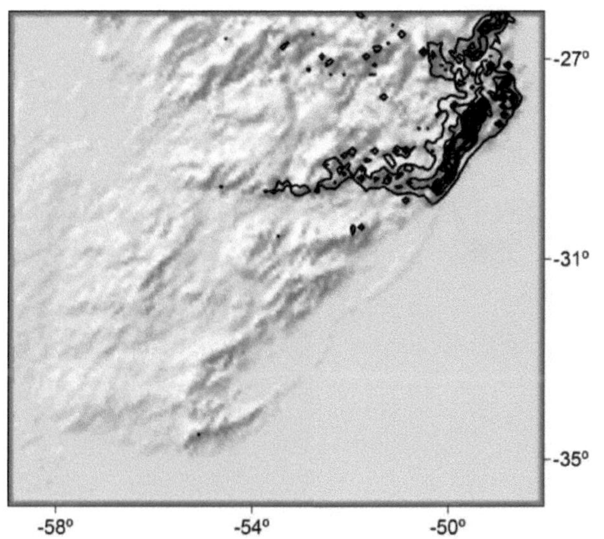

Figura 7 - Spatial layout of the TC calculated by TC2DFTPL for the ASTER MDT with 10 km resolution

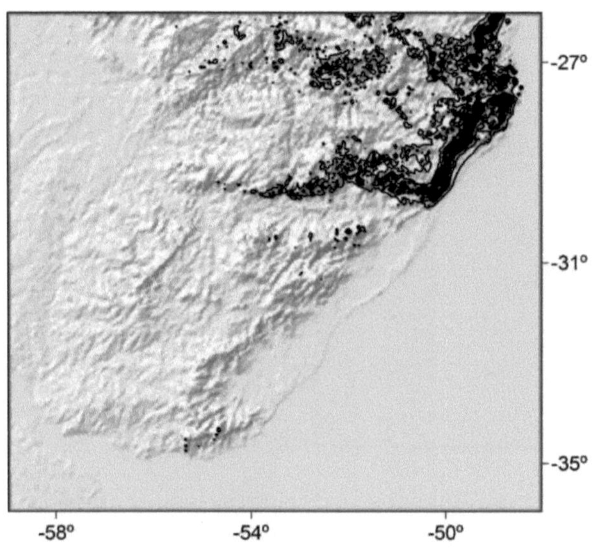

Figura 8 - Spatial layout of the TC calculated by TC2DFTPL for the ASTER MDT with 5 km resolution

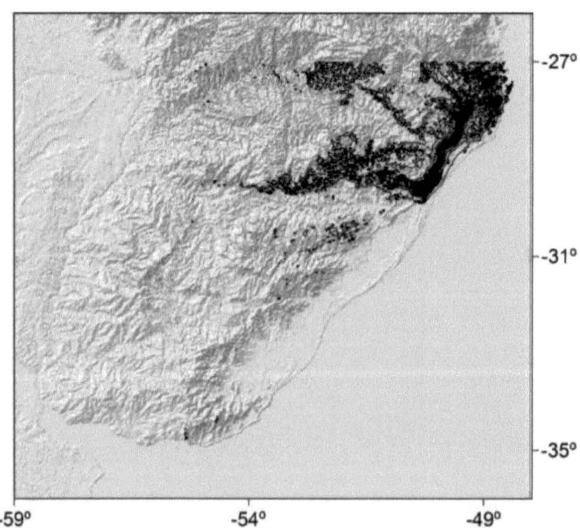

Figura 9 - Spatial layout of the TC calculated by TC2DFTPL for the ASTER MDT

with 2 km resolution

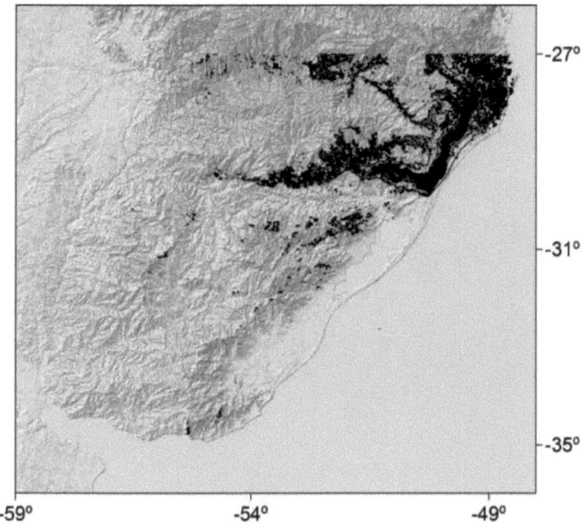

Figura 10 - Spatial layout of the TC calculated by TC2DFTPL for the ASTER MDT with 1 km resolution

Figures 11 to 14 show the spatial layout of the terrain correction calculated by the TCFOUR programme from the SRTM digital elevation models at different resolutions, and Figures 15 to 18 show the spatial layout of the terrain correction calculated by the same programme from the ASTER digital elevation models at different resolutions.

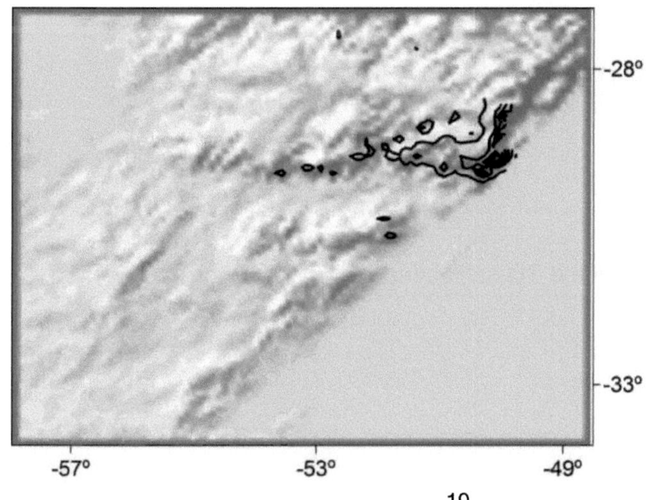

Figure 11 - Spatial layout of the TC calculated by TCFOUR for the SRTM MDT with 10 km resolution

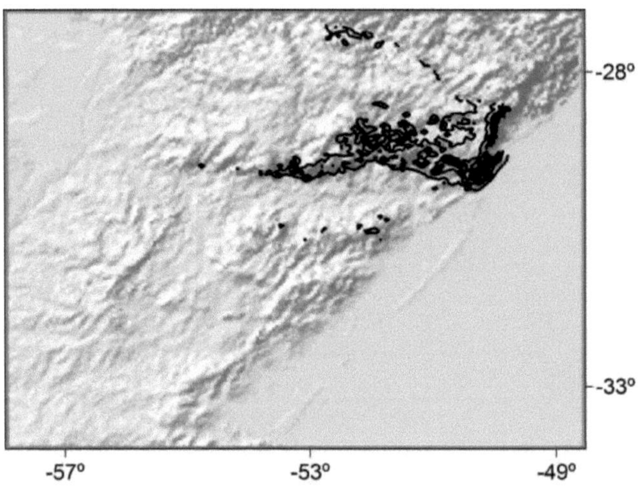

Figura 12 - Spatial layout of the TC calculated by TCFOUR for the SRTM MDT with 5 km resolution

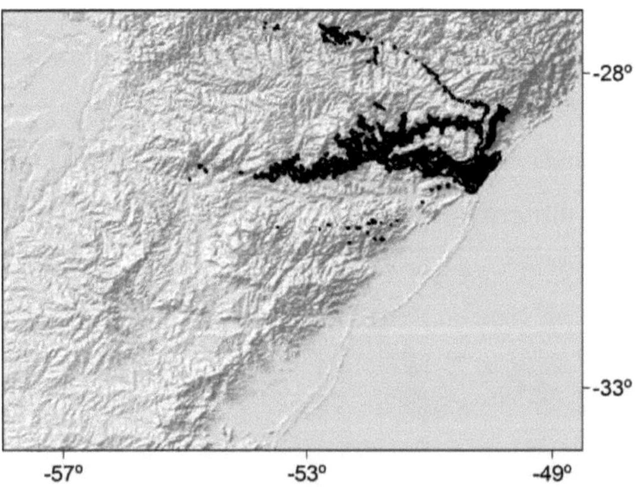

Figura 13 - Spatial layout of the TC calculated by TCFOUR for the SRTM MDT with 2 km resolution

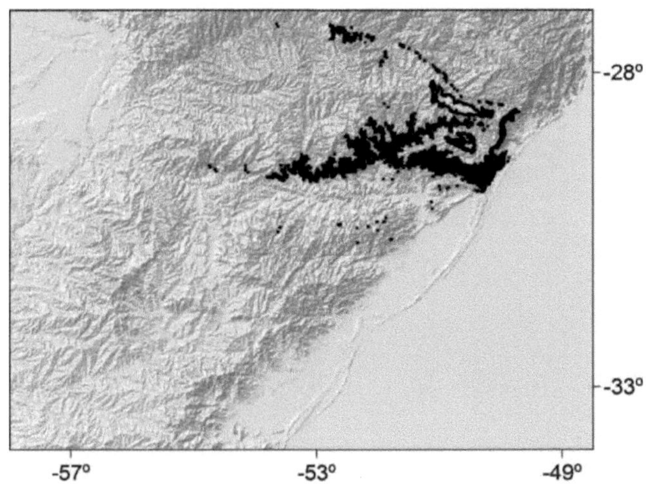

Figura 14 - Spatial layout of the TC calculated by TCFOUR for the SRTM MDT with 1 km resolution

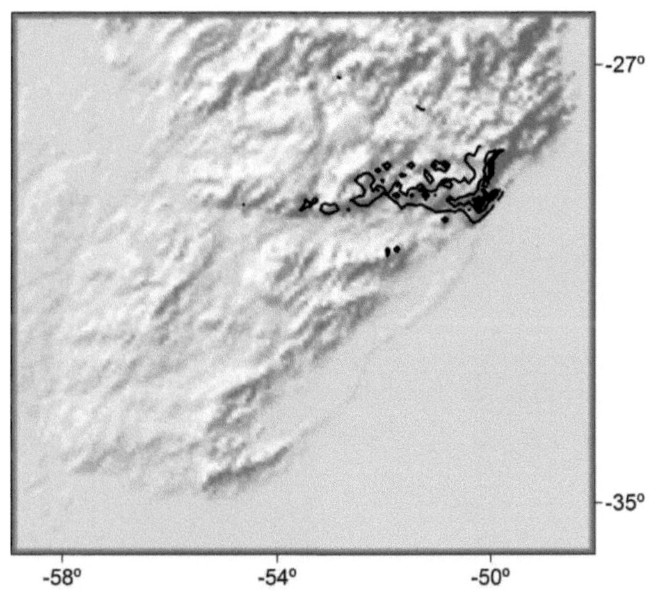

Figura 15 - Spatial layout of the TC calculated by TCFOUR for the ASTER MDT with 10 km resolution

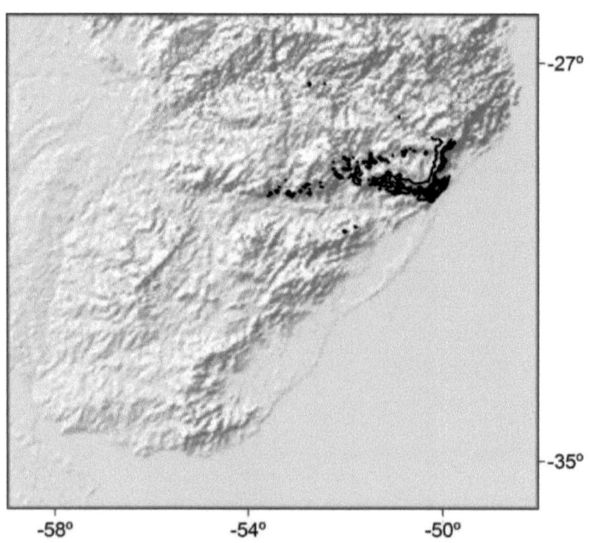

Figura 16 - Spatial layout of the TC calculated by TCFOUR for the ASTER MDT with 5 km resolution

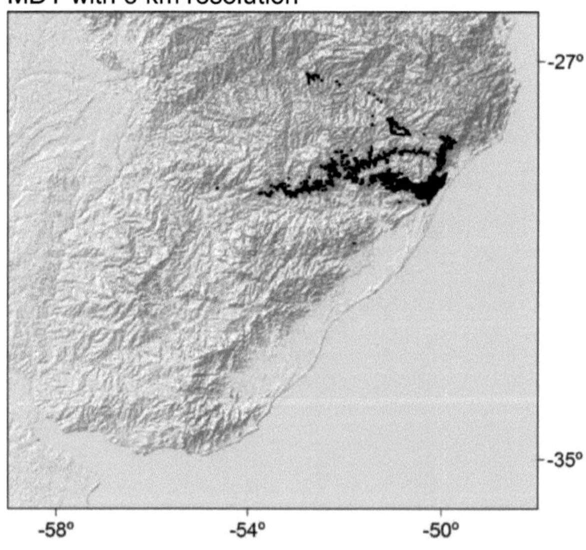

Figura 17 - Spatial layout of the TC calculated by TCFOUR for the ASTER MDT with 2 km resolution

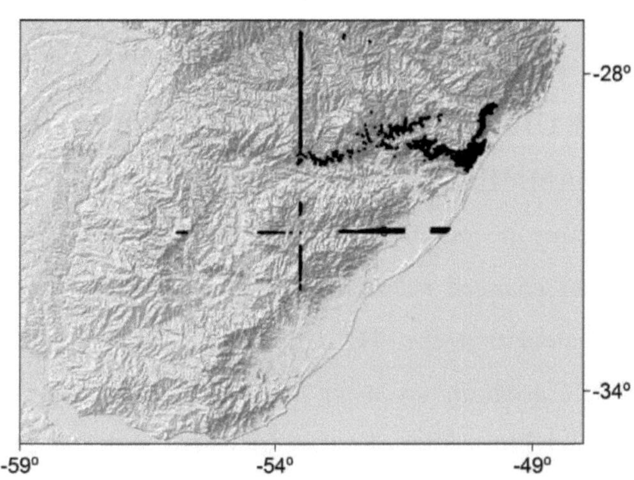

Figura 18 - Spatial layout of the TC calculated by TCFOUR for the ASTER MDT with 1 km resolution

CHAPTER 5

COMPARISON OF THE CT CALCULATED BY THE TC2DFTPL AND TCFOUR PROGRAMMES

In order to check whether the region of RS where the terrain correction takes on the highest values changed according to the use of a different resolution MDT when calculating the TC, the TC obtained by the TC2DFTPL and TCFOUR programmes using the SRTM and ASTER MDTs at 5 km, 2 km and 1 km resolutions was compared with the TC obtained by these programmes using the MDTs at 10 km resolution.

Figures 19, 20 and 21 show, respectively, the comparison (represented by points) between the TC calculated by the TC2DFTPL programme using the SRTM MDTs at 5 km, 2 km and 1 km resolutions with the TC calculated by this programme using the 10 km MDT.

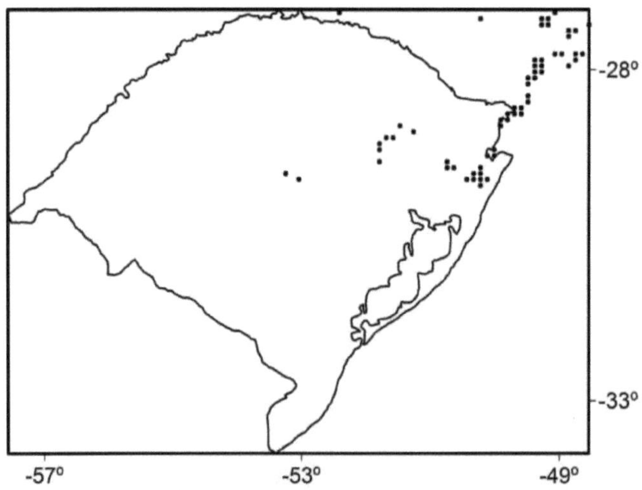

Figure 19 - TC2DFTPL_SRTM: 5 km x 10 km

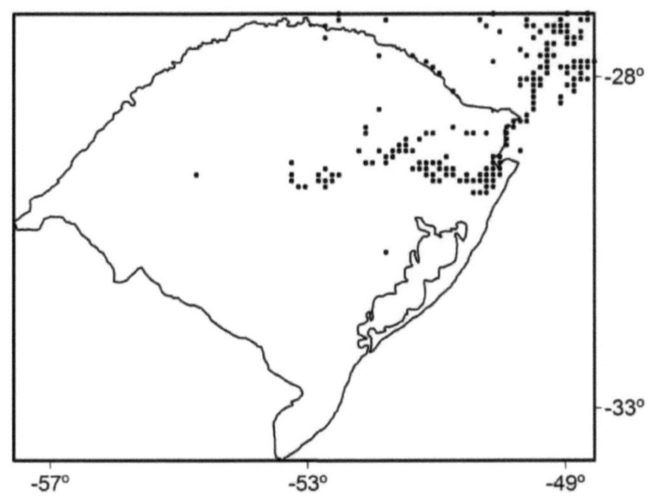

Figure 20 - TC2DFTPL_SRTM: 2 km x 10 km

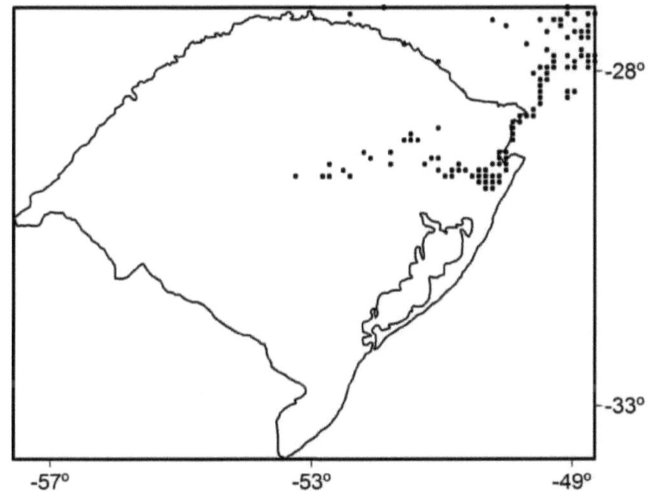

Figure 21 - TC2DFTPL_SRTM: 1 km x 10 km

Figures 22, 23 and 24 show, respectively, the comparison (represented by points) between the TC calculated by the TCFOUR programme using the SRTM MDTs at 5 km, 2 km and 1 km resolutions with the TC calculated by this programme using the 10 km MDT.

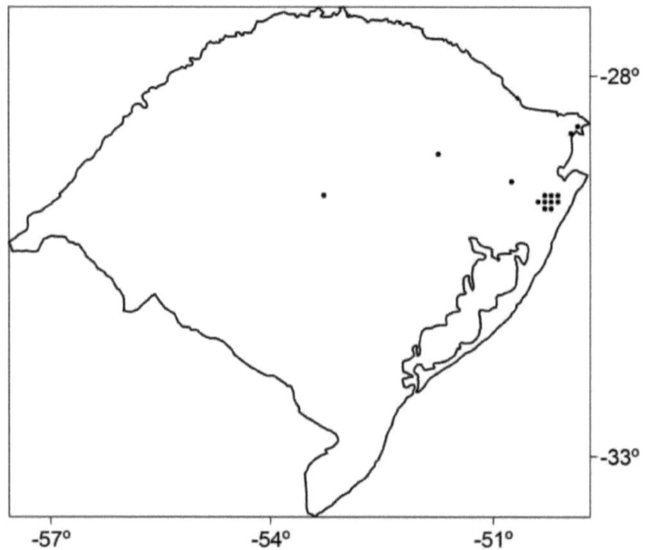

Figure 22 - TCFOUR_SRTM: 5 km x 10 km

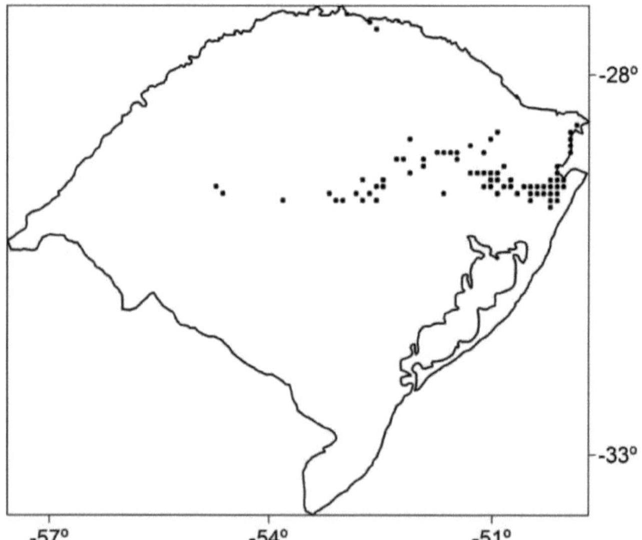

Figure 23 - TCFOUR_SRTM: 2 km x 10 km

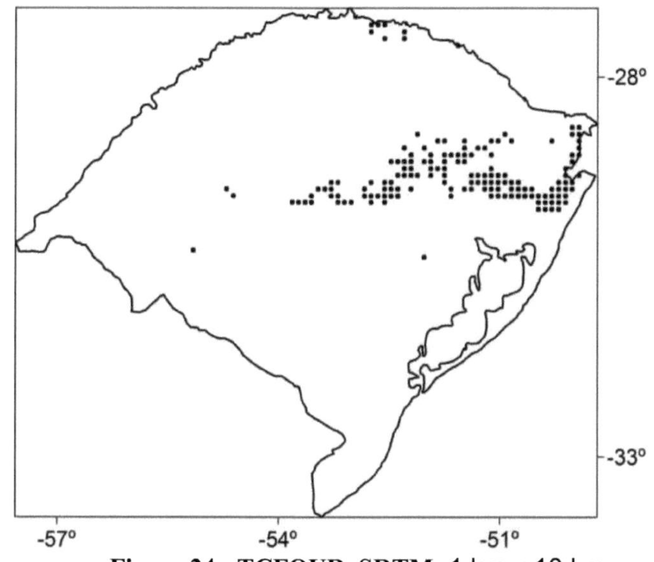

Figure 24 - TCFOUR_SRTM: 1 km x 10 km

The interpretation of Figures 19 to 24 shows that the differences (represented by dots) greater than 0.9 mGal (in module) between the terrain corrections for the lowest resolution model (10 km) and the highest resolution models (5 km, 2 km and 1 km) are concentrated in the region of RS with the highest altitudes in the state, known as Serra Geral.

The CTs calculated with the ASTER MDTs were compared in the same way and the results were similar to those obtained with the SRTM MDTs (Figures 25 to 30). It was observed that the differences greater than 0.9 mGal (in module) between the TCs calculated for the lowest resolution model (10 km) and the highest resolution models (5 km, 2 km and 1 km) are also concentrated in the Serra Geral region and the edge effect can be seen in the subdivisions made in the study area for the 1 km model. The dots in Figure 30 represent the difference between the TCs calculated using the ASTER MDTs with resolutions of 1 km and 10 km and, circled in black, the edge effect observed in the divisions of the study area.

27

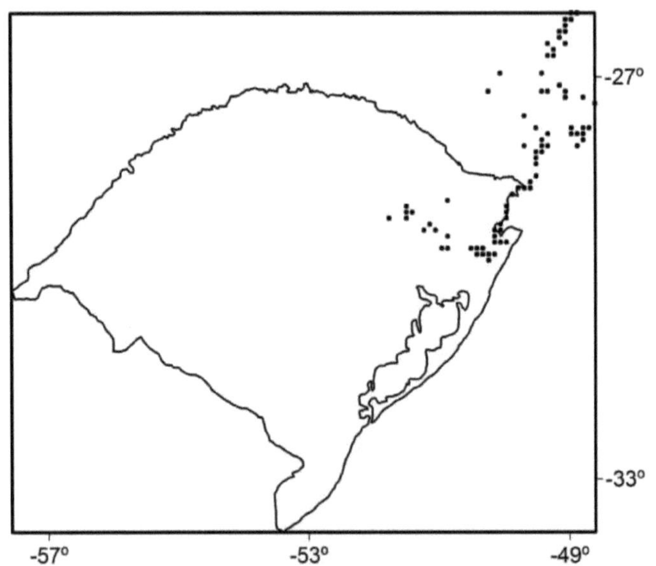

Figure 25 - TC2DFTPL_ASTER: 5 km x 10 km

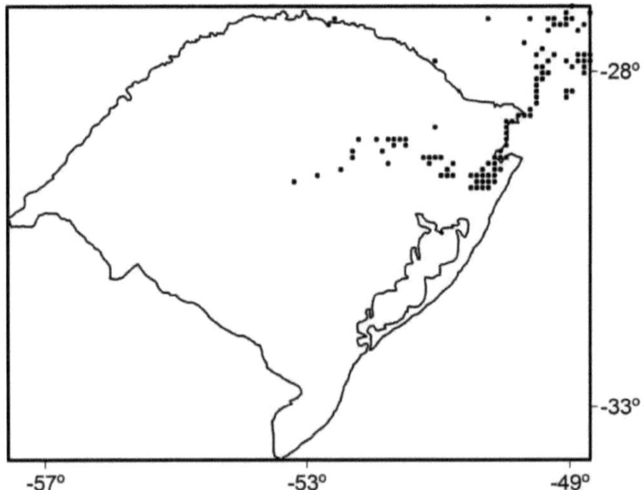

Figure 26 - TC2DFTPL_ASTER: 2 km x 10 km

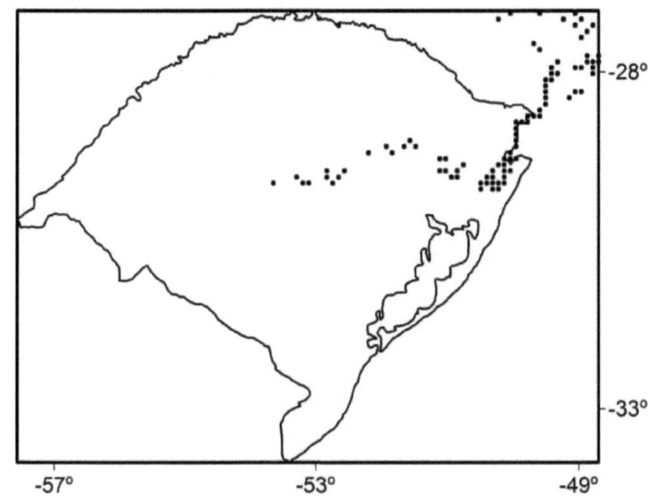

Figure 27 - TC2DFTPL_ASTER: 1 km x 10 km

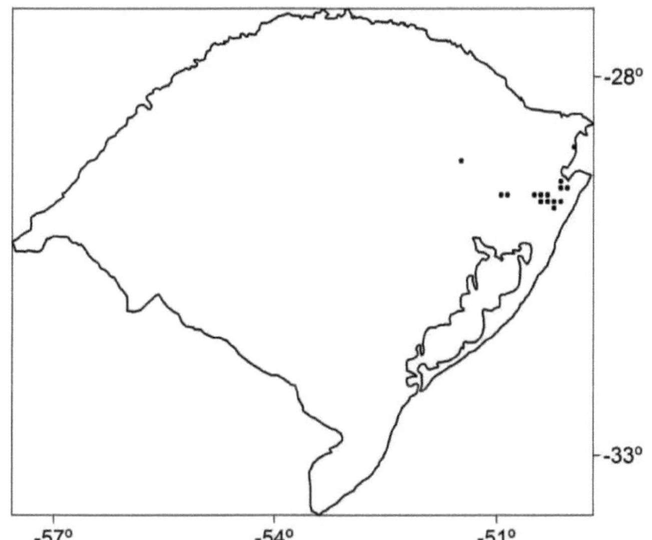

Figure 28 - TCFOUR_ASTER: 5 km x 10 km

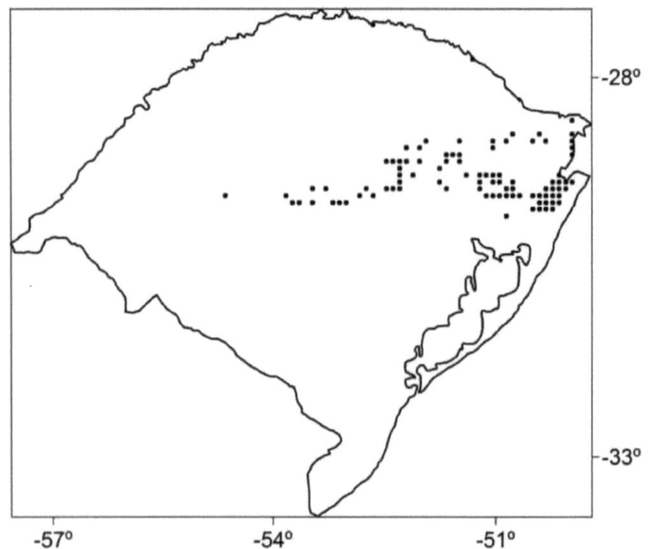

Figure 29 - TCFOUR_ASTER: 2 km x 10 km

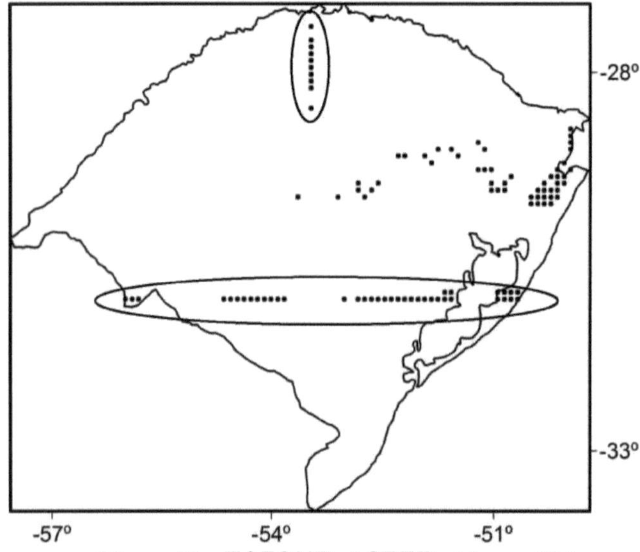

Figure 30 - TCFOUR_ASTER: 1 km x 10 km

The TC calculated by both programmes was then compared using the SRTM MDTs at 10 km, 5 km, 2 km and 1 km resolutions. The terrain corrections calculated by the two programmes for the 10 km model are similar, as no difference of more than 1 mGal (in module) was identified.

The points in Figures 31, 32 and 33 represent the differences between the TCs calculated by the two programmes. An analysis of Figure 31 shows that the TCs calculated by the two programmes for the 5 km model are also similar, as the differences greater than 1 mGal (in module) appear in small quantities and are concentrated on the north-eastern edge of the RS.

However, analysing Figures 32 and 33, it can be seen that the differences greater than 1 mGal (in module) between the 2 km and 1 km models appear in greater quantities than those of the 5 km model and are concentrated in the Serra Geral region of RS.

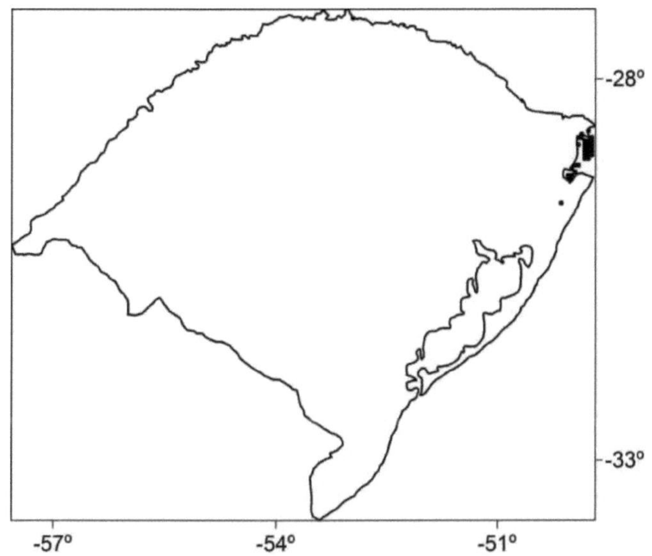

Figure 31 - TC2DFTPL x TCFOUR_SRTM: 5 km x 5 km

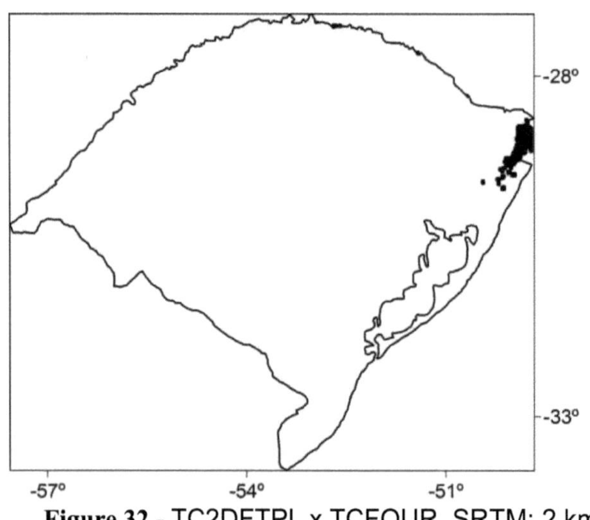

Figure 32 - TC2DFTPL x TCFOUR_SRTM: 2 km x 2 km

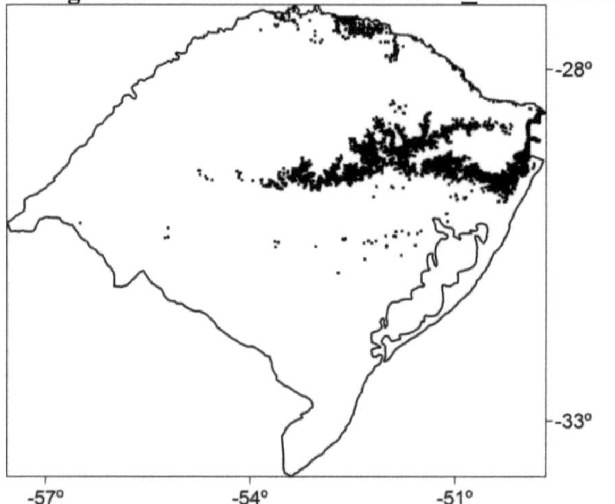

Figure 33 - TC2DFTPL x TCFOUR_SRTM: 1 km x 1 km

Just as was found using the SRTM MDTs, the TCs calculated by the two programmes using the ASTER MDTs for the 10 km model are similar, as no difference greater than 1 mGal (modulo) was identified and the differences greater than 1 mGal (modulo) between the 5 km, 2 km and 1 km models are also concentrated in the Serra Geral region, as can be seen in Figures 34 to 36. Figure 36 shows the differences (represented by dots)

32

between the TCs calculated by both programmes for the 1 km ASTER model and, circled in black, the edge effect observed in the divisions made in the study area.

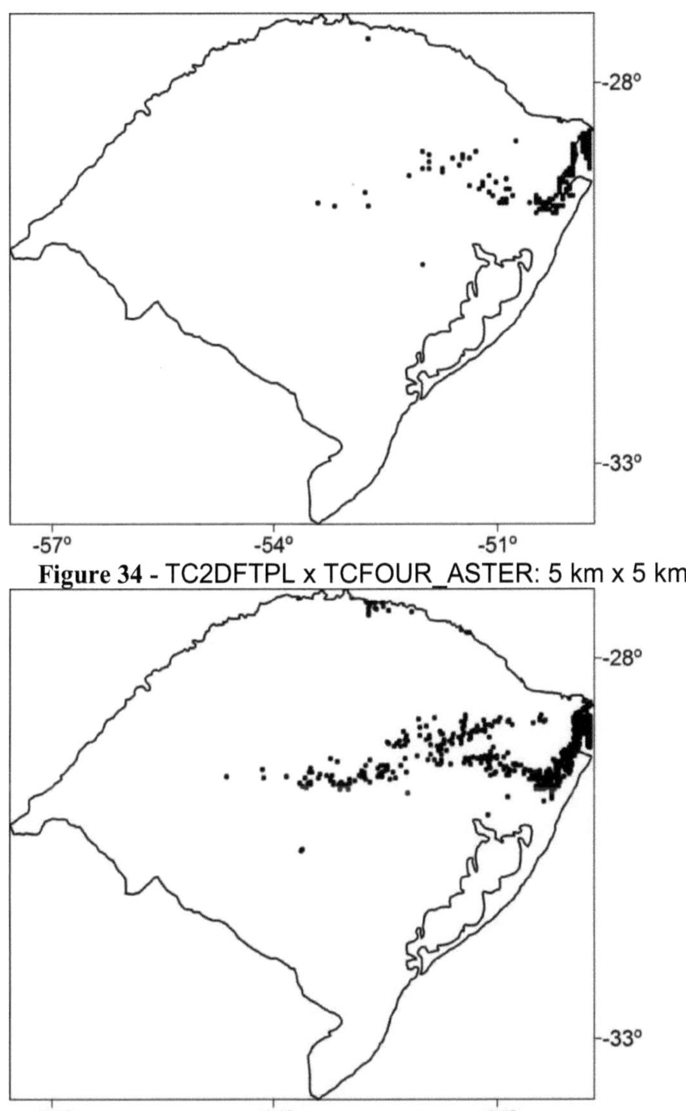

Figure 34 - TC2DFTPL x TCFOUR_ASTER: 5 km x 5 km

Figure 35 - TC2DFTPL x TCFOUR_ASTER: 2 km x 2 km

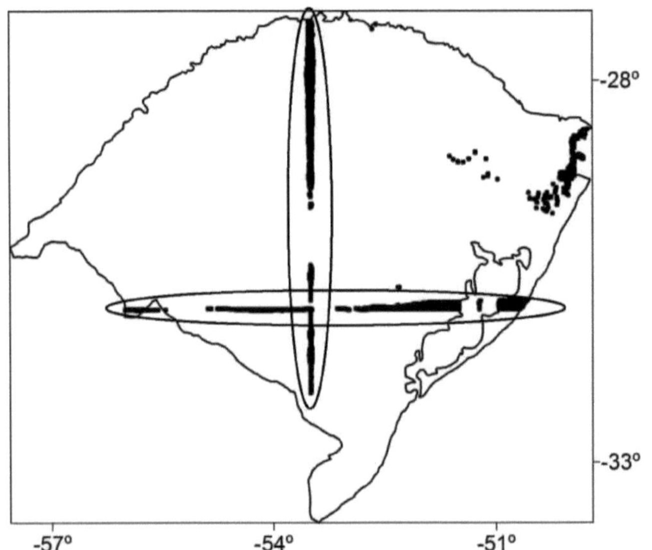

Figure 36 - TC2DFTPL x TCFOUR_ASTER: 1 km x 1 km

CHAPTER 6

RESULTS AND DISCUSSIONS

The terrain correction statistics calculated by the TC2DFTPL and TCFOUR programmes using SRTM and ASTER data can be seen in Tables 2, 3, 4 and 5.

Table 2 - TC statistics calculated by TCFOUR with SRTM MDTs.

SRTM - TCFOUR				
Statistics (mGal)	**MDT resolution**			
	10 kilometres	**5 kilometres**	**2 kilometres**	**1 kilometre**
Minimum	0,0124	0,0130	0,0150	-0,1792
Maximum	3,2260	6,4740	12,3951	17,6582
Average	0,1529	0,1971	0,2831	0,3271
Standard Deviation	0,2466	0,3468	0,5459	0,6975

Table 3 - TC statistics calculated by TCFOUR with ASTER MDTs.

ASTER - TCFOUR				
Statistics (mGal)	**MDT resolution**			
	10 kilometres	**5 kilometres**	**2 kilometres**	**1 kilometre**
Minimum	0,0120	0,0130	-0,0730	0,0010
Maximum	3,5070	6,0780	13,8210	11,6769
Average	0,1595	0,2048	0,2893	0,3024
Standard Deviation	0,2571	0,3772	0,5927	0,5963

Analysing Tables 2 and 3, it can be seen that the TC calculated by the TCFOUR programme using the SRTM MDT with a resolution of 1 km and using the ASTER MDT with a resolution of 2 km, has an inconsistent (negative) minimum value. In addition, the TC statistics for the 1 km resolution ASTER MDT (Table 3) calculated by the TCFOUR programme show compromised values due to the presence of the edge effect in the divisions of the study area (Figure 36).

Tables 4 and 5 show that the TC calculated by the TC2DFTPL programme using the SRTM and ASTER MDTs does not show negative values, as observed in the TC calculated by the TCFOUR programme (Tables 2 and 3). However, the minimum and maximum values for the SRTM MDT at 1 km resolution (Table 4) and for the ASTER MDT at 2 km and 1 km resolutions (Table 5) are underestimated, due to the need to compartmentalise the study area for these resolutions.

Table 4 - CT statistics calculated by TC2DFTPL with SRTM MDTs.

SRTM - TC2DFTPL

Statistics (mGal)	10 kilometres	MDT resolution		
		5 kilometres	2 kilometres	1 kilometre
Minimum	0,0020	0,0020	0,0020	0
Maximum	7,0610	10,9790	20,1380	13,1037
Average	0,1160	0,1522	0,2120	0,1715
Standard Deviation	0,3050	0,4252	0,6106	0,4981

Table 5 - CT statistics calculated by TC2DFTPL with ASTER MDTs.

ASTER - TC2DFTPL

Statistics (mGal)	10 kilometres	MDT resolution		
		5 kilometres	2 kilometres	1 kilometre
Minimum	0,0010	0,0010	0,0010	0
Maximum	6,438	10,4510	10,7467	13,2143

| Average | 0,1004 | 0,1302 | 0,1402 | 0,1200 |
| Standard Deviation | 0,2697 | 0,3868 | 0,4531 | 0,4154 |

For the purpose of comparing the terrain correction statistics calculated by the two programmes, the edge effect was disregarded. It can thus be seen that there is a directly proportional relationship between the resolution of the digital terrain model used to calculate the terrain correction and the value of the correction to be applied to the gravimetric data: as the resolution of the DEM used increases, the terrain correction to be applied to the gravimetric data also increases. This is because the denser the DEM, the greater the amount of information you have about the study area and, therefore, the TC value increases. However, this is not the case when it is necessary to divide up the study area (see Table 1) so that the programmes can calculate the TC. Therefore, in cases where it is necessary to divide the study area, the value of the terrain correction calculated will be underestimated (especially the corrections calculated by the TC2DFTPL programme, as can be seen in Tables 4 and 5). When the study area is divided up, the TC is underestimated due to the programme not being able to process all the data at once. Thus, the greater the amount of data, the greater the smoothing in determining the TC. Tables 6 and 7 show, respectively, the statistics relating to the comparison between the TCs calculated by the TC2DFTPL and TCFOUR programmes for the differences found greater than 1 mGal (in module), for the SRTM and ASTER MDTs at 10 km, 5 km, 2 km and 1 km resolutions.

Table 6 - CT comparison statistics calculated with the SRTM MDTs.

SRTM - TC2DFTPL x TCFOUR
MDT resolution

Statistics (mGal)	10 kilometres	5 kilometres	2 kilometres	1 kilometre
Minimum	-0,9314	-2,7299	-7,9367	-2,9363
Maximum	0,4700	1,6162	6,0104	11,3941
Average	0,0010	0,2553	0,8731	1,8465
Standard Deviation	0,0471	1,5334	2,2456	1,1058

Table 7 - CT comparison statistics calculated with the ASTER MDTs.

	ASTER - TC2DFTPL x TCFOUR			
Statistics (mGal)	10 kilometres	MDT resolution		
		5 kilometres	2 kilometres	1 kilometre
Minimum	-0,7390	1,0010	-5,4507	-5,6386
Maximum	0,4126	4,7716	5,8827	11,6750
Average	-0,0046	1,6959	1,7415	2,3872
Standard Deviation	0,0425	0,6404	1,3191	1,9927

Tables 6 and 7 show that for the 10 km model, for both the SRTM and ASTER MDTs, no differences greater than 1 mGal (in module) were found between the TCs calculated by both programmes. However, for the other resolutions (5 km, 2 km and 1 km) differences greater than 1 mGal (modulo) were found, appearing in greater quantities in the 2 km and 1 km resolutions (as seen in Figures 32 and 33) and in smaller quantities in the 5 km resolution (as seen in Figure 31).

CHAPTER 7

CONCLUSIONS

At the end of the evaluations, it can be concluded that the TC2DFTPL and TCFOUR programmes for calculating terrain correction are similar and that the TC values calculated using MDTs with a resolution of 5 km will be underestimated, since for higher resolutions (2 km and 1 km) both programmes do not detect the improvement of the MDT. This is considered a problem when the aim is to determine local geoids.

However, we recommend using the TC2DFTPL programme to calculate the terrain correction. This is because the TC2DFTPL programme has no edge effect, unlike the TCFOUR programme. However, care should be taken with its use when the study area is very large and needs to be compartmentalised; as can be seen in Tables 4 and 5, the results presented will be underestimated due to the need to divide up the study area.

Tables 6 and 7 also show that the results of the comparison statistics between the maximum and minimum values obtained for the terrain correction calculated by both programmes using the SRTM and ASTER MDTs at 5 km, 2 km and 1 km resolutions are due to the need to divide the study area for the higher resolutions and also to the altimetric differences between the two MDTs, which are more evident in the region of higher altitudes in the study area, such as the Serra Geral in the state of Rio Grande do Sul.

CHAPTER 8

REFERENCES

BLITZKOW, D.; MATOS, A. C. O. C.; CINTRA, J. P. SRTM evaluation in Brazil and Argentina with emphasis on the Amazon region. International Association of Geodesy Symposia. v.130, p. 266-271, 2007.

DRUZINA, A. G. S. Integration of altimetric data obtained through different techniques to generate a new digital elevation model. Porto Alegre, 2007. 79p. Master's dissertation in Remote Sensing, State Centre for Remote Sensing and Meteorology Research, Federal University of Rio Grande do Sul.

FORSBERG, R. A study of terrain reductions, density anomalies and geophysical inversion methods in gravity field modelling. Scientific Report N. 5, p. 33-69, 1984.

FORSBERG, R.; TSCHERNING, C. C. An overview manual for the GRAVSOFT. Geodetic Gravity Field Modelling Programs. Denmark, 2008. 2ed. 75p.

GEMAEL, C. Introduction to Physical Geodesy. Curitiba. UFPR, 2002. 304p.

GISAT. 2014. ASTER GDEM Readme File. Available at: <http://www.gisat.cz/content/en/products/digital-elevation-model/aster-gdem>. Accessed on 06/01/2014.

HAMMER, S. Terrain corrections for gravimeter stations. Geophysics, v.4, p. 184-194, 1939.

IBGE States. 2014. Rio Grande do Sul. Available at: <http://www.ibge.gov.br/estadosat/perfil.php?sigla=rs>. Accessed on 17/12/2014.

JAMUR, K. P.; PEREIRA, R. A. D.; de FREITAS, S. R. C.; MIRANDA, F. D. A.; FERREIRA, V. G. Evaluation of the gravimetric terrain correction obtained from digital elevation models associated with the SGB and EGM2008. III Brazilian Symposium on Geodetic Sciences and

Geoinformation Technologies. p. 001-005, 2010.

JAMUR, K. P.; de FREITAS, S. R. C. Study of alternatives for combining satellite altimetry and gravimetry for use in regions with low conventional gravimetric coverage. III Brazilian Geomatics Symposium. p. 280-295, 2012.

LEMOS, M. C.; SOUZA, S. F.; ROCHA, R. S. Evaluation of the quality of altimetric data derived from the Shuttle Radar Topographic Mission (SRTM): Preliminary Results. Proceedings of the 1st Symposium on Geodetic Sciences and Geoinformation Technologies. Recife, 2004. Available at: <http://www.ufpe.br/cgtg/ISIMGEO/CD/html/Fotogrametria%20e%20Sen soriamento%20Remoto/Artigos/f008.pdf>. Accessed on 26/11/2013.

LI, Y. C.; SIDERIS, M. G. Improved gravimetric terrain corrections. Geophysical Journal International, v. 119, p. 740-752, 1994.

LOBIANCO, M. C. B. Determination of geoid heights in Brazil. São Paulo, 2005. 165p, p. 61-69. Doctoral Thesis in Engineering, Polytechnic School, University of São Paulo.

NASA. 2013. Shuttle Radar Topography Mission. Available at: <http://www2.jpl.nasa.gov/srtm/>. Accessed on 24/01/2013.

MATOS, A. C. O. C. Implementation of digital terrain models for applications in Geodesy and Geophysics in South America. São Paulo, 2005. 355p. Doctoral Thesis in Engineering, Polytechnic School, University of São Paulo.

RODRIGUEZ, E.; MORRIS, C. S.; BELZ, J. E.; CHAPIN, E. C.; MARTIN, J. M.; DAFFER, W.; HENSLEY, S. An assessment of the SRTM topographic products. Technical Report JPL D-31639, Jet Propulsion Laboratory. Pasadena, California, 2005. 143p.

SIDERIS, M. G. A fast fourier transform method of computing terrain corrections. Manuscripta Geodaetica, v. 10, p. 66-73, 1985.

Buy your books fast and straightforward online - at one of world's fastest growing online book stores! Environmentally sound due to Print-on-Demand technologies.

Buy your books online at
www.morebooks.shop

Kaufen Sie Ihre Bücher schnell und unkompliziert online – auf einer der am schnellsten wachsenden Buchhandelsplattformen weltweit! Dank Print-On-Demand umwelt- und ressourcenschonend produziert.

Bücher schneller online kaufen
www.morebooks.shop

Printed by Books on Demand GmbH, Norderstedt / Germany